FREE Test Taking Tips DVD Offer

To help us better serve you, we have developed a Test Taking Tips DVD that we would like to give you for FREE. **This DVD covers world-class test taking tips that you can use to be even more successful when you are taking your test.**

All that we ask is that you email us your feedback about your study guide. Please let us know what you thought about it – whether that is good, bad or indifferent.

To get your **FREE Test Taking Tips DVD**, email freedvd@studyguideteam.com with "FREE DVD" in the subject line and the following information in the body of the email:

  a. The title of your study guide.

  b. Your product rating on a scale of 1-5, with 5 being the highest rating.

  c. Your feedback about the study guide. What did you think of it?

  d. Your full name and shipping address to send your free DVD.

If you have any questions or concerns, please don't hesitate to contact us at freedvd@studyguideteam.com.

Thanks again!

Grade 3 Math Workbook

Math Workbooks Grade 3 Team

Table of Contents

Chapter One: Basic Numbers

Even and Odd Numbers

Are these numbers even or odd?
Write your answer in the blank.

1. 7 _____

2. 3 _____

3. 88 _____

4. 2 _____

5. 1 _____

6. 55 _____

7. 11 _____

8. 448 _____

9. 711 _____

10. 540 _____

11. 87 _____

12. 139 _____

13. 99 _____

14. 326 _____

15. 258 _____

16. 90 _____

17. 237 _____

18. 19 _____

19. 100 _____

20. 77 _____

Rounding to the nearest 100

Round the following numbers to the nearest hundred.
Write your answer in the blank.

1. 246 _____

2. 761 _____

3. 72 _____

4. 388 _____

5. 550 _____

6. 142 _____

7. 870 _____

8. 261 _____

9. 652 _____

10. 734 _____

11. 667 _____

12. 851 _____

13. 348 _____

14. 169 _____

15. 508 _____

16. 456 _____

17. 239 _____

18. 105 _____

19. 152 _____

20. 613 _____

Rounding to the nearest 1,000

Round the following numbers to the nearest thousand.
Write your answer in the blank.

1. 5,767 _____   11. 5,627 _____

2.   789 _____   12. 4,502 _____

3. 3,323 _____   13. 2,386 _____

4. 8,199 _____   14. 3,620 _____

5.   552 _____   15. 8,011 _____

6. 9,471 _____   16. 6,228 _____

7. 2,186 _____   17. 1,425 _____

8. 7,545 _____   18. 4,916 _____

9. 6,439 _____   19. 1,020 _____

10. 1,880 _____   20. 2,567 _____

Place Value by 10

Count the groups inside the boxes,
then write the numbers beside them.
The first problem has been finished for you as an example.

O = 1   ⬜⬜⬜⬜⬜⬜⬜⬜⬜⬜ = 10

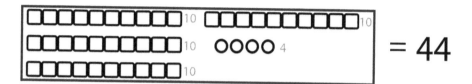

= 44

1.    =

2. 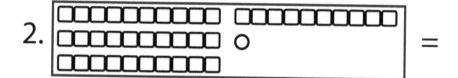   =

3. 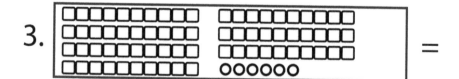   =

4. 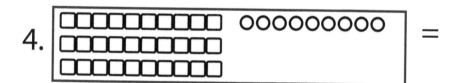   =

5.   =

Place Value by 100

Count the groups inside the boxes,
then write the numbers beside them.
The first problem has been finished for you as an example.

○ = 1   ⬚⬚⬚⬚⬚⬚⬚⬚⬚⬚ = 10   ▦ = 100

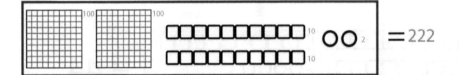 = 222

1. 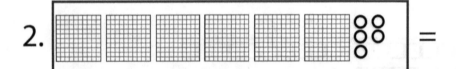 =

2. =

3. =

4. =

5. =

Place Value by 1,000
Count the groups inside the boxes,
then write the numbers beside them.
The first problem has been finished for you as an example.

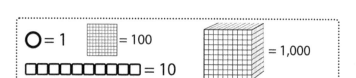

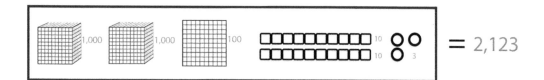 $= 2,123$

1. 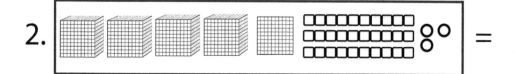 $=$

2. 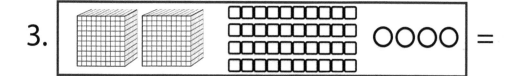 $=$

3. 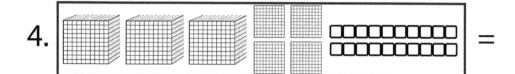 $=$

4. $=$

5. $=$

Least to Greatest

Write these numbers from least to greatest in the blank.

1. 567 · 289 · 233 – _____

2. 1,894 · 2,002 · 1,770 – _____

3. 4,200 · 4,188 · 4,006 – _____

4. 8,090 · 7,612 · 8,211 – _____

5. 909 · 3,201 · 1,222 – _____

6. 7,233 · 6,997 · 7,431 – _____

7. 5,023 · 5025 · 5,199 – _____

8. 3,229 · 3,139 · 3,073 – _____

9. 4,862 · 4,928 · 3,634 – _____

10. 99 · 108 · 2,078 – _____

Comparing Numbers

Compare the numbers.
Write <, >, or = in the blank box for each question

1. 79 ☐ 98

2. 9,121 ☐ 9,120

3. 3,713 ☐ 3,713

4. 205 ☐ 29

5. 6,994 ☐ 7,151

6. 2,003 ☐ 274

7. 5,482 ☐ 5,487

8. 2,102 ☐ 2,097

9. 1,305 ☐ 1,310

10. 204 ☐ 208

11. 1,310 ☐ 1,310

12. 8,928 ☐ 8,910

13. 7,818 ☐ 7,862

14. 5,000 ☐ 4,911

15. 7,110 ☐ 6,810

16. 1,477 ☐ 1,442

Answers: Chapter 1

Even and Odd Numbers

1. Odd	11. Odd
2. Odd	12. Odd
3. Even	13. Odd
4. Even	14. Even
5. Odd	15. Even
6. Odd	16. Even
7. Odd	17. Odd
8. Even	18. Odd
9. Odd	19. Even
10. Even	20. Odd

Rounding to the 100

1. 200	11. 700
2. 800	12. 900
3. 100	13. 300
4. 400	14. 200
5. 600	15. 500
6. 100	16. 500
7. 900	17. 200
8. 300	18. 100
9. 700	19. 200
10. 700	20. 600

Rounding to the 1,000

1. 6,000	11. 6,000
2. 1,000	12. 5,000
3. 3,000	13. 2,000
4. 8,000	14. 4,000
5. 1,000	15. 8,000
6. 9,000	16. 6,000
7. 2,000	17. 1,000
8. 8,000	18. 5,000
9. 6,000	19. 1,000
10. 2,000	20. 3,000

Place Value by 10

1. 13
2. 41
3. 76
4. 39
5. 95

Answers: Chapter 1

Place Value by 100

1. 231
2. 605
3. 136
4. 823
5. 366

Place Value by 1,000

1. 1,326
2. 4,133
3. 2,044
4. 3,420
5. 5,208

Least to Greatest

1. 233 · 289 · 567
2. 1,770 · 1,894 · 2,002
3. 4,006 · 4,188 · 4,200
4. 7,612 · 8,090 · 8,211
5. 909 · 1,222 · 3,201
6. 6,997 · 7,233 · 7,431
7. 5,023 · 5,025 · 5,199
8. 3,073 · 3,139 · 3,229
9. 3,634 · 4,862 · 4,928
10. 99 · 108 · 2,078

Comparing Numbers

1. <
2. >
3. =
4. >
5. <
6. >
7. <
8. >
9. <
10. <
11. =
12. >
13. <
14. >
15. >
16. >

12

Chapter Two: Addition

Addition

2-Digit Addition: A

1.
```
  87
+ 12
____
```

2.
```
  93
+ 46
____
```

3.
```
  44
+ 54
____
```

4.
```
  70
+ 22
____
```

5.
```
  18
+ 99
____
```

6.
```
  10
+ 75
____
```

7.
```
  16
+ 33
____
```

8.
```
  24
+ 69
____
```

9.
```
  58
+ 76
____
```

10.
```
  88
+ 85
____
```

11.
```
  11
+ 41
____
```

12.
```
  17
+ 35
____
```

13.
```
  37
+ 90
____
```

14.
```
  22
+ 77
____
```

15.
```
  15
+ 28
____
```

16.
```
  47
+ 82
____
```

17.
```
  59
+ 39
____
```

18.
```
  12
+ 94
____
```

19.
```
  55
+ 56
____
```

20.
```
  98
+ 96
____
```

Addition

2-Digit Addition: B

1.
```
   61
 + 81
_____
```

2.
```
   70
 + 18
_____
```

3.
```
   38
 + 34
_____
```

4.
```
   29
 + 47
_____
```

5.
```
   39
 + 91
_____
```

6.
```
   62
 + 14
_____
```

7.
```
   96
 + 78
_____
```

8.
```
   48
 + 49
_____
```

9.
```
   88
 + 13
_____
```

10.
```
   23
 + 47
_____
```

11.
```
   50
 + 80
_____
```

12.
```
   11
 + 54
_____
```

13.
```
   20
 + 68
_____
```

14.
```
   55
 + 18
_____
```

15.
```
   38
 + 77
_____
```

16.
```
   90
 + 90
_____
```

17.
```
   66
 + 34
_____
```

18.
```
   18
 + 52
_____
```

19.
```
   43
 + 58
_____
```

20.
```
   39
 + 29
_____
```

Addition
3-Digit Addition: A

1.
```
   483
 + 999
_____
```

2.
```
   369
 + 721
_____
```

3.
```
   963
 + 331
_____
```

4.
```
   587
 + 147
_____
```

5.
```
   298
 + 522
_____
```

6.
```
   852
 + 677
_____
```

7.
```
   719
 + 928
_____
```

8.
```
   211
 + 376
_____
```

9.
```
   890
 + 154
_____
```

10.
```
   396
 + 613
_____
```

11.
```
   790
 + 800
_____
```

12.
```
   183
 + 920
_____
```

13.
```
   487
 + 564
_____
```

14.
```
   234
 + 268
_____
```

15.
```
   102
 + 129
_____
```

16.
```
   521
 + 814
_____
```

17.
```
   746
 + 725
_____
```

18.
```
   352
 + 386
_____
```

19.
```
   259
 + 390
_____
```

20.
```
   455
 + 412
_____
```

Addition

3-Digit Addition: B

1.
$$\begin{array}{r} 711 \\ + 288 \\ \hline \end{array}$$

2.
$$\begin{array}{r} 505 \\ + 900 \\ \hline \end{array}$$

3.
$$\begin{array}{r} 670 \\ + 172 \\ \hline \end{array}$$

4.
$$\begin{array}{r} 210 \\ + 440 \\ \hline \end{array}$$

5.
$$\begin{array}{r} 328 \\ + 800 \\ \hline \end{array}$$

6.
$$\begin{array}{r} 720 \\ + 859 \\ \hline \end{array}$$

7.
$$\begin{array}{r} 551 \\ + 277 \\ \hline \end{array}$$

8.
$$\begin{array}{r} 110 \\ + 344 \\ \hline \end{array}$$

9.
$$\begin{array}{r} 201 \\ + 374 \\ \hline \end{array}$$

10.
$$\begin{array}{r} 637 \\ + 619 \\ \hline \end{array}$$

11.
$$\begin{array}{r} 429 \\ + 500 \\ \hline \end{array}$$

12.
$$\begin{array}{r} 942 \\ + 237 \\ \hline \end{array}$$

13.
$$\begin{array}{r} 850 \\ + 930 \\ \hline \end{array}$$

14.
$$\begin{array}{r} 756 \\ + 115 \\ \hline \end{array}$$

15.
$$\begin{array}{r} 519 \\ + 713 \\ \hline \end{array}$$

16.
$$\begin{array}{r} 325 \\ + 852 \\ \hline \end{array}$$

17.
$$\begin{array}{r} 222 \\ + 489 \\ \hline \end{array}$$

18.
$$\begin{array}{r} 369 \\ + 900 \\ \hline \end{array}$$

19.
$$\begin{array}{r} 423 \\ + 472 \\ \hline \end{array}$$

20.
$$\begin{array}{r} 580 \\ + 518 \\ \hline \end{array}$$

Addition

4-Digit Addition

1.
   1125
 + 4739
 _____

2.
   3894
 + 9666
 _____

3.
   8473
 + 8778
 _____

4.
   2947
 + 3829
 _____

5.
   4700
 + 3782
 _____

6.
   2904
 + 5829
 _____

7.
   6393
 + 5899
 _____

8.
   7444
 + 4719
 _____

9.
   9822
 + 9153
 _____

10.
   3899
 + 4947
 _____

11.
   6233
 + 1382
 _____

12.
   1270
 + 1194
 _____

13.
   5222
 + 3929
 _____

14.
   6293
 + 2983
 _____

15.
   3000
 + 1740
 _____

16.
   2392
 + 5829
 _____

17.
   8003
 + 7002
 _____

18.
   4700
 + 5829
 _____

19.
   9377
 + 8573
 _____

20.
   2556
 + 4628
 _____

Addition Squares

1. Add the numbers going down.
2. Then add the numbers going across.
3. Finally, add together the sums to find the answer.
The first problem has been finished for you as an example.

9	5	14
11	6	17
20	11	31

1.

5	6	
3	4	

2.

47	3	
83	9	

3.

12	1	
1	12	

4.

7	3	
9	4	

5.

10	48	
23	4	

6.

78	12	
45	2	

7.

25	53	
28	9	

Answers: Chapter 2

2-Digit Addition: A

1. 99	11. 52
2. 139	12. 52
3. 98	13. 127
4. 92	14. 99
5. 117	15. 43
6. 85	16. 129
7. 49	17. 98
8. 93	18. 106
9. 134	19. 111
10. 173	20. 194

3-Digit Addition: A

1. 1482	11. 1590
2. 1090	12. 1103
3. 1294	13. 1051
4. 734	14. 502
5. 820	15. 231
6. 1529	16. 1335
7. 1647	17. 1471
8. 587	18. 738
9. 1044	19. 649
10. 1009	20. 867

2-Digit Addition: B

1. 142	11. 130
2. 88	12. 65
3. 72	13. 88
4. 76	14. 73
5. 130	15. 115
6. 76	16. 180
7. 174	17. 100
8. 97	18. 70
9. 101	19. 101
10. 70	20. 68

3-Digit Addition: B

1. 999	11. 929
2. 1405	12. 1179
3. 842	13. 1780
4. 650	14. 871
5. 1128	15. 1232
6. 1579	16. 1177
7. 828	17. 711
8. 454	18. 1269
9. 575	19. 895
10. 1256	20. 1098

Answers: Chapter 2

4-Digit Addition

1. 5864
2. 13,560
3. 17,251
4. 6776
5. 8482
6. 8733
7. 12,292
8. 12,163
9. 18,975
10. 8846

11. 7615
12. 2464
13. 9151
14. 9276
15. 4740
16. 8221
17. 15,005
18. 10,529
19. 17,950
20. 7184

Addition Squares

1.

5	6	11
3	4	7
8	10	18

4.

7	3	10
9	4	13
16	7	23

7.

25	53	78
28	9	37
53	62	115

2.

47	3	50
83	9	92
130	12	142

5.

10	48	58
23	4	27
33	52	85

3.

12	1	13
1	12	13
13	13	26

6.

78	12	90
45	2	47
123	14	137

Chapter Three: Subtraction

Subtraction

2-Digit Subtraction: A

1.
```
   98
 - 23
```

2.
```
   47
 - 17
```

3.
```
   54
 - 30
```

4.
```
   81
 - 40
```

5.
```
   72
 - 60
```

6.
```
   26
 - 22
```

7.
```
   37
 - 18
```

8.
```
   88
 - 50
```

9.
```
   59
 - 12
```

10.
```
   67
 - 29
```

11.
```
   86
 - 19
```

12.
```
   75
 - 35
```

13.
```
   99
 - 17
```

14.
```
   87
 - 79
```

15.
```
   38
 - 11
```

16.
```
   96
 - 38
```

17.
```
   83
 - 29
```

18.
```
   55
 - 36
```

19.
```
   44
 - 24
```

20.
```
   73
 - 67
```

Subtraction

2-Digit Subtraction: B

1.
```
    14
  - 11
  _____
```

2.
```
    93
  - 55
  _____
```

3.
```
    56
  - 35
  _____
```

4.
```
    97
  - 12
  _____
```

5.
```
    36
  - 16
  _____
```

6.
```
    81
  - 47
  _____
```

7.
```
    73
  - 24
  _____
```

8.
```
    75
  - 39
  _____
```

9.
```
    57
  - 11
  _____
```

10.
```
    25
  - 16
  _____
```

11.
```
    95
  - 51
  _____
```

12.
```
    87
  - 10
  _____
```

13.
```
    52
  - 48
  _____
```

14.
```
    45
  - 21
  _____
```

15.
```
    69
  - 48
  _____
```

16.
```
    28
  - 12
  _____
```

17.
```
    91
  - 30
  _____
```

18.
```
    54
  - 32
  _____
```

19.
```
    37
  - 16
  _____
```

20.
```
    71
  - 66
  _____
```

Subtraction
3-Digit Subtraction: A

1.
```
  988
- 147
```

2.
```
  247
- 239
```

3.
```
  560
- 358
```

4.
```
  839
- 471
```

5.
```
  690
- 121
```

6.
```
  535
- 200
```

7.
```
  199
- 158
```

8.
```
  772
- 385
```

9.
```
  975
- 844
```

10.
```
  310
- 266
```

11.
```
  783
- 288
```

12.
```
  273
- 215
```

13.
```
  430
- 388
```

14.
```
  900
- 374
```

15.
```
  705
- 533
```

16.
```
  838
- 655
```

17.
```
  540
- 190
```

18.
```
  626
- 498
```

19.
```
  947
- 698
```

20.
```
  801
- 108
```

Subtraction

3-Digit Subtraction: B

1.
```
  374
- 255
```

2.
```
  974
- 327
```

3.
```
  800
- 466
```

4.
```
  588
- 299
```

5.
```
  847
- 127
```

6.
```
  690
- 555
```

7.
```
  384
-310
```

8.
```
  440
- 368
```

9.
```
  901
- 101
```

10.
```
  720
- 264
```

11.
```
  700
- 299
```

12.
```
  188
- 103
```

13.
```
  611
- 430
```

14.
```
  427
- 284
```

15.
```
  274
- 177
```

16.
```
  526
- 488
```

17.
```
  944
- 188
```

18.
```
  808
- 288
```

19.
```
  756
- 100
```

20.
```
  611
- 399
```

27

Subtraction

4-Digit Subtraction

1.
```
  3883
- 2999
_____
```

2.
```
  9004
- 8362
_____
```

3.
```
  5626
- 4288
_____
```

4.
```
  8463
- 2884
_____
```

5.
```
  9836
- 1001
_____
```

6.
```
  1847
- 1233
_____
```

7.
```
  6720
- 4966
_____
```

8.
```
  4739
- 2995
_____
```

9.
```
  2797
- 1475
_____
```

10.
```
  6739
- 4000
_____
```

11.
```
  7649
- 6284
_____
```

12.
```
  1948
- 1007
_____
```

13.
```
  2589
- 2398
_____
```

14.
```
  5200
- 4288
_____
```

15.
```
  9462
- 4759
_____
```

16.
```
  3447
- 2999
_____
```

17.
```
  8305
- 3305
_____
```

18.
```
  6622
- 3000
_____
```

19.
```
  4029
- 2199
_____
```

20.
```
  7800
- 3888
_____
```

Subtraction

Fill in the blank with the missing number.

1. ___ - 40 = 12

2. 102 - ___ = 74

3. ___ - 13 = 225

4. 78 - ___ = 21

5. ___ - 38 = 89

6. 74 - ___ = 29

7. ___ - 18 = 95

8. ___ - 7 = 176

9. ___ - 89 = 37

10. ___ - 109 = 2

11. 17 - ___ = 6

12. 59 - ___ = 13

13. 96 - ___ = 29

14. 192 - ___ = 47

15. 73 - ___ = 4

16. ___ - 64 = 15

17. ___ - 198 = 223

18. ___ - 37 = 199

19. 142 - ___ = 101

20. 208 - ___ = 5

Answers: Chapter 3

2-Digit Subtraction: A

1. 75	11. 67
2. 30	12. 40
3. 24	13. 82
4. 41	14. 8
5. 12	15. 27
6. 4	16. 58
7. 19	17. 54
8. 38	18. 19
9. 47	19. 20
10. 38	20. 6

3-Digit Subtraction: A

1. 841	11. 495
2. 8	12. 58
3. 202	13. 42
4. 368	14. 526
5. 569	15. 172
6. 335	16. 183
7. 41	17. 350
8. 387	18. 128
9. 131	19. 249
10. 44	20. 693

2-Digit Subtraction: B

1. 3	11. 44
2. 38	12. 77
3. 21	13. 4
4. 85	14. 24
5. 20	15. 21
6. 34	16. 16
7. 49	17. 61
8. 36	18. 22
9. 46	19. 21
10. 9	20. 5

3-Digit Subtraction: B

1. 119	11. 401
2. 647	12. 85
3. 334	13. 181
4. 289	14. 143
5. 720	15. 97
6. 135	16. 38
7. 74	17. 756
8. 72	18. 520
9. 800	19. 656
10. 456	20. 212

Answers: Chapter 3

4-Digit Subtraction

1. 884	11. 1365
2. 642	12. 941
3. 1338	13. 191
4. 5579	14. 912
5. 8835	15. 4703
6. 614	16. 448
7. 1754	17. 5000
8. 1744	18. 3622
9. 1322	19. 1830
10. 2739	20. 3912

Subtraction: Fill in the Blanks

1. 52	11. 11
2. 28	12. 46
3. 238	13. 67
4. 57	14. 145
5. 127	15. 69
6. 45	16. 79
7. 113	17. 421
8. 183	18. 236
9. 126	19. 41
10. 111	20. 203

Chapter Four: Multiplication

Multiplication

Break each group down by writing
the problem out, then adding it together.
The first problem has been finished for you as an example.

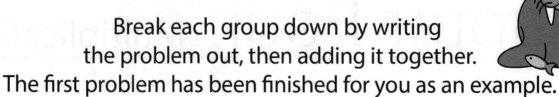

$$4 \times 7 =$$
$$\underline{7} + \underline{7} + \underline{7} + \underline{7} = 28$$

1. 2×8

2. 3×6

3. 2×9

4. 4×5

5. 2×3

1. 4×7

6. 3×8

7. 5×3

8. 4×9

9. 5×6

Multiplication
Single Digit Multiplication

1.
```
    3
  x 4
_____
```

2.
```
    2
  x 8
_____
```

3.
```
    6
  x 6
_____
```

4.
```
    5
  x 3
_____
```

5.
```
    8
  x 7
_____
```

6.
```
    6
  x 4
_____
```

7.
```
    3
  x 7
_____
```

8.
```
    7
  x 6
_____
```

9.
```
    9
  x 6
_____
```

10.
```
    5
  x 8
_____
```

11.
```
    4
  x 7
_____
```

12.
```
    3
  x 3
_____
```

13.
```
    2
  x 7
_____
```

14.
```
    6
  x 8
_____
```

15.
```
    4
  x 4
_____
```

16.
```
    7
  x 9
_____
```

17.
```
    8
  x 3
_____
```

18.
```
    3
  x 5
_____
```

19.
```
    3
  x 2
_____
```

20.
```
    4
  x 5
_____
```

Multiplication

Fill in the blank with the missing number

1. __ x 5 = 30

2. 4 x __ = 12

3. __ x 8 = 40

4. 7 x __ = 28

5. 9 x __ = 18

6. __ x 7 = 42

7. __ x 9 = 36

8. 5 x __ = 30

9. __ x 2 = 16

10. 3 x __ = 12

Multiplication
Single Digit Multiplication

1.
```
    3
  x 4
  ___
```

2.
```
    2
  x 8
  ___
```

3.
```
    6
  x 6
  ___
```

4.
```
    5
  x 3
  ___
```

5.
```
    8
  x 7
  ___
```

6.
```
    6
  x 4
  ___
```

7.
```
    3
  x 7
  ___
```

8.
```
    7
  x 6
  ___
```

9.
```
    9
  x 6
  ___
```

10.
```
    5
  x 8
  ___
```

11.
```
    4
  x 7
  ___
```

12.
```
    3
  x 3
  ___
```

13.
```
    2
  x 7
  ___
```

14.
```
    6
  x 8
  ___
```

15.
```
    4
  x 4
  ___
```

16.
```
    7
  x 9
  ___
```

17.
```
    8
  x 3
  ___
```

18.
```
    3
  x 5
  ___
```

19.
```
    3
  x 2
  ___
```

20.
```
    4
  x 5
  ___
```

Multiplication

Multiplication by 100

1.
```
  100
x   4
_____
```

2.
```
  100
x   3
_____
```

3.
```
  100
x   8
_____
```

4.
```
  100
x   9
_____
```

5.
```
  100
x   1
_____
```

6.
```
  100
x   2
_____
```

7.
```
  100
x   5
_____
```

8.
```
  100
x   6
_____
```

9.
```
  100
x   7
_____
```

10.
```
    3
x 100
_____
```

11.
```
    2
x 100
_____
```

12.
```
    4
x 100
_____
```

13.
```
    8
x 100
_____
```

14.
```
    7
x 100
_____
```

15.
```
    9
x 100
_____
```

Multiplication
2-Digit Multiplication

1.
```
    12
  x  2
_____
```

2.
```
    50
  x  2
_____
```

3.
```
    30
  x  3
_____
```

4.
```
    16
  x  2
_____
```

5.
```
    15
  x  3
_____
```

6.
```
    25
  x  3
_____
```

7.
```
    18
  x  2
_____
```

8.
```
    42
  x  2
_____
```

9.
```
    37
  x  2
_____
```

10.
```
    14
  x  3
_____
```

11.
```
    17
  x  3
_____
```

12.
```
    24
  x  2
_____
```

13.
```
    56
  x  1
_____
```

14.
```
    33
  x  3
_____
```

15.
```
    19
  x  2
_____
```

Answers: Chapter 4

Multiplication Rules

1. 8 + 8 = 16
2. 6 + 6 + 6 = 18
3. 9 + 9 = 18
4. 5 + 5 + 5 + 5 = 20
5. 3 + 3 = 6
6. 7 + 7 + 7 + 7 = 28
7. 8 + 8 + 8 = 24
8. 3 + 3 + 3 + 3 + 3 = 15
9. 9 + 9 + 9 + 9 = 36
10. 6 + 6 + 6 + 6 + 6 = 30

Fill in the Blank

1. 6
2. 3
3. 5
4. 4
5. 2
6. 6
7. 4
8. 6
9. 8
10. 4

Single Digit Multiplication

1. 12
2. 16
3. 36
4. 15
5. 56
6. 24
7. 21
8. 42
9. 54
10. 40
11. 28
12. 9
13. 14
14. 48
15. 16
16. 63
17. 24
18. 15
19. 6
20. 20

Multiplication by 10

1. 80
2. 90
3. 70
4. 20
5. 70
6. 60
7. 30
8. 10
9. 50
10. 40
11. 30
12. 10
13. 90
14. 70
15. 50

Answers: Chapter 4

Multiplication by 100

1. 400
2. 300
3. 800
4. 900
5. 100
6. 200
7. 500
8. 600
9. 700
10. 300

11. 200
12. 400
13. 800
14. 700
15. 900

2-Digit Multiplication

1. 24
2. 100
3. 90
4. 32
5. 45
6. 75
7. 36
8. 84
9. 74
10. 42

11. 51
12. 48
13. 56
14. 99
15. 38

Chapter Five: Division

Division

Divide the items evenly by the animals in each group.
The first problem has been finished for you as an example.

Object	Animals	Answer
		3
1.		
2.		
3.		
4.		
5.		

Division

Divide each group by the numbers in each box and write the answers
The first problem has been finished for you as an example.

	Object	Divided By	Answers
	(pencils)	÷ 2	6
		÷ 3	4
		÷ 6	2
1.	(basketballs)	÷ 3	—
		÷ 7	—
			—
2.	(dogs)	÷ 2	—
		÷ 4	—
		÷ 8	—
3.	(stars)	÷ 3	—
		÷ 5	—
4.	(triangles)	÷ 4	—
		÷ 8	—
5.	(airplanes)	÷ 2	—
		÷ 4	—
		÷ 8	—

Division Wording and Terms

Complete each sentence.
The first problem has been finished for you as an example.

50/5 = 10 is read: " _50_ divided by _5_ is equal to _10_."

1. 16/8 = 2 is read: "____ divided by ____ is equal to ____."

2. 28/7 = 4 is read : "____ divided by ____ is equal to ____."

3. 32/4 = 8 is read : "____ divided by ____ is equal to ____."

4. 48/8 = 6 is read : "____ divided by ____ is equal to ____."

5. 15/5 = 3 is read : "____ divided by ____ is equal to ____."

56/7 = 8. The divisor is _56_, the dividend is _7_, the quotient is _8_.

6.   56/7 = 8. The divisor is ____, the dividend is ____, the quotient is ____.

7. 50/5 = 10. The divisor is ____, the dividend is ____, the quotient is ____.

8.   18/9 = 2. The divisor is ____, the dividend is ____, the quotient is ____.

9.   36/4 = 9. The divisor is ____, the dividend is ____, the quotient is ____.

10. 42/7 = 6. The divisor is ____, the dividend is ____, the quotient is ____.

Dividing by 10 and 100

Use division to answer the following questions.

1. $80 \div 10 =$

2. $1,000 \div 100 =$

3. $9,430 \div 10 =$

4. $6,750 \div 10 =$

5. $910 \div 10 =$

6. $510 \div 10 =$

7. $7,640 \div 100 =$

8. $420 \div 10 =$

9. $5,100 \div 100 =$

10. $4,650 \div 10 =$

11. $370 \div 10 =$

12. $830 \div 10 =$

13. $2,600 \div 10 =$

14. $9,500 \div 100 =$

15. $5,600 \div 100 =$

16. $490 \div 10 =$

17. $390 \div 10 =$

18. $2,400 \div 10 =$

19. $1,340 \div 10 =$

20. $750 \div 10 =$

Division Word Problems

Write out each problem and solve it.

1. Cindy has 28 crayons and wants to divide them into equal groups. Show two ways she can do that.

_____   _____

2. Ricardo has 60 lollipops and wants to divide them into equal groups. Show two ways he can do that.

_____   _____

3. Benji has 16 books and wants to divide them into equal groups. Show two ways he can do that.

_____   _____

4. Lucy has 24 muffins and wants to divide them into equal groups. Show two ways she can do that.

_____   _____

5. Peter has 30 flowers and wants to divide them into equal groups. Show two ways she can do that.

_____

Basic division

Solve each problem .

1.

$9 \overline{)81}$

2.

$8 \overline{)64}$

3.

$7 \overline{)56}$

4.

$3 \overline{)24}$

5.

$9 \overline{)54}$

6.

$4 \overline{)16}$

7.

$6 \overline{)18}$

8.

$8 \overline{)32}$

9.

$3 \overline{)12}$

10.

$7 \overline{)49}$

11.

$6 \overline{)48}$

12.

$5 \overline{)25}$

13.

$4 \overline{)36}$

14.

$21 \overline{)7}$

15.

$2 \overline{)18}$

Dividing Objects

A	B
1. 2	1. 7, 3
2. 4	2. 8, 4, 2
3. 4	3. 5, 3
4. 3	4. 6, 3
5. 2	5. 12, 8, 4

Wording and Terms

1. "16 divided by 8 is equal to 2."

2. "28 divided by 7 is equal to 4."

3. "32 divided by 4 is equal to 8."

4. "48 divided by 8 is equal to 6."

5. "15 divided by 5 is equal to 3."

6. The divisor is 56, the dividend is 7, the quotient is 8.

7. The divisor is 50, the dividend is 5, the quotient is 10.

8. The divisor is 18, the dividend is 9, the quotient is 2.

9. The divisor is 36, the dividend is 4, the quotient is 9.

10. The divisor is 42, the dividend is 7, the quotient is 6.

Answers: Chapter 5

Dividing by 10 and 100

1. 8	11. 37
2. 10	12. 83
3. 943	13. 260
4. 675	14. 95
5. 91	15. 56
6. 51	16. 49
7. 764	17. 39
8. 42	18. 240
9. 51	19. 134
10. 465	20. 75

Basic Division

1. 9	11. 8
2. 8	12. 5
3. 8	13. 9
4. 8	14. 3
5. 6	15. 9
6. 4	
7. 3	
8. 4	
9. 4	
10. 7	

Word Problems

There are more than two solutions to each
problem, but a few are listed below.

1. $28 \div 7$, $28 \div 4$, $28 \div 2$

2. $60 \div 10$, $60 \div 6$, $60 \div 20$, $60 \div 15$

3. $16 \div 8$, $16 \div 4$, $16 \div 2$

4. $24 \div 8$, $24 \div 3$, $24 \div 6$, $24 \div 2$

5. $30 \div 6$, $30 \div 5$, $30 \div 10$, $30 \div 3$

Chapter Six: Fractions

Fractions With Shapes

Color ½ of each shape.

1.

2.

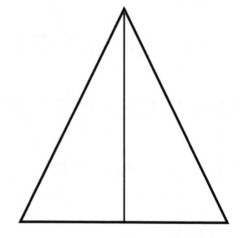

3.

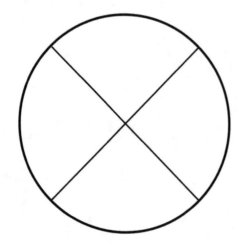

4.

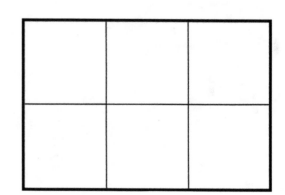

5.

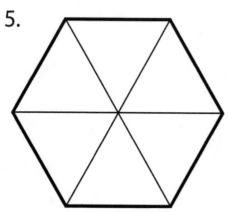

Fractions With Shapes

Color ⅓ of each shape.

1.

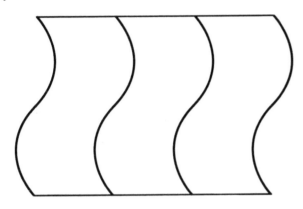

2.

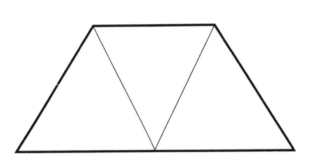

3.

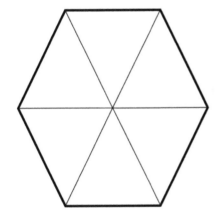

4.

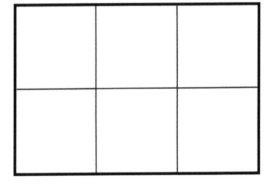

5.

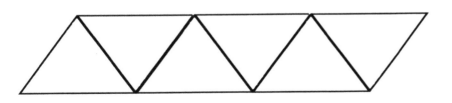

Fractions With Shapes

Color ¼ of each shape.

1.

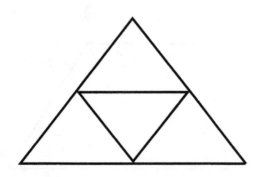

2.

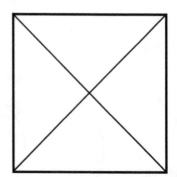

3.

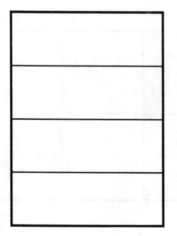

4.

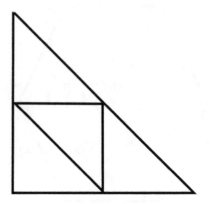

5.

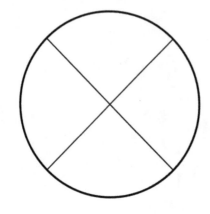

Identifying Fractions

Write what fraction of each set is shaded in.

1.

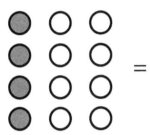   =

2.

=

3.

=

4.

=

Color each set to match the fraction.

5.

9/15 or 3/5 =

6.

6/24 or 1/4 =

7.

5/30 or 1/6 =

8.

14/42 or 2/7 =

57

Matching Fractions

Match each Picture (on the left) with the correct Fraction (on the right).
Write your answer in the blank next to the picture.

1.

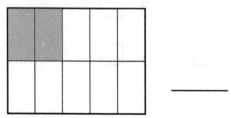

_____

2.

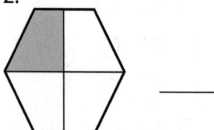

_____

3.

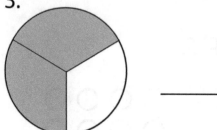

_____

4.

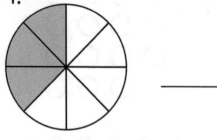

_____

5.

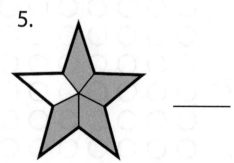

_____

A. 2/3

B. 2/10 or 1/5

C. 4/5

D. 1/4

E. 3/8

Whole Numbers and Fractions

Find the missing number to complete the fraction.

1. $1 = \dfrac{}{4}$

2. $1 = \dfrac{}{8}$

3. $1 = \dfrac{}{16}$

4. $1 = \dfrac{}{2}$

5. $1 = \dfrac{}{6}$

Write the fraction that is equal to the whole number.

6. $35 =$

7. $62 =$

8. $9 =$

9. $12 =$

10. $24 =$

Answers: Chapter 6

Fractions with Shapes

1/2	1/3	1/4

1.

1.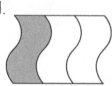

1.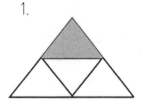

2.

2.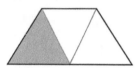

2.

3.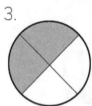

3.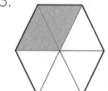

3.

4.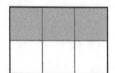

4.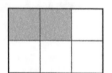

4.

5.

5.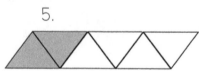

5.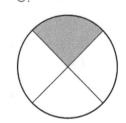

Answers: Chapter 6

Fractions from Objects

1. 4/12 or 1/3

2. 10/15 or 1/3

3. 7/16

4. 4/20 or 1/5

5.

6.

7.

8.

Matching Fractions

1. B

2. D

3. A

4. E

5. C

Whole Numbers

1. $\dfrac{4}{4}$

2. $\dfrac{8}{8}$

3. $\dfrac{16}{16}$

4. $\dfrac{2}{2}$

5. $\dfrac{6}{6}$

6. $\dfrac{35}{1}$

7. $\dfrac{62}{1}$

8. $\dfrac{9}{1}$

9. $\dfrac{12}{1}$

10. $\dfrac{24}{1}$

Chapter Seven: Money and Time

Coin Value

Write the value for each coin pictured below.

1.

= _____

2.

= _____

3.

= _____

4.

= _____

Dollar Value

Write the value for each bill pictured below.

1.

= _____

2.

= _____

3.

= _____

4.

= _____

Counting Coins

Count the value for each group of coins.

1.   = _____

2.   = _____

3.   = _____

4.   = _____

5.   = _____

6.    = _____

7.   = _____

8.   = _____

Counting Dollars

Count the value for each group of bills.

1.

    = _____

2.

   = _____

3.

    = _____

4.

    = _____

Counting Dollars

Count the value for each group of bills.

5.

= _____

6.

= _____

7.

= _____

8.

= _____

Counting Dollars and Coins

Write the correct amount of money for each question.

1.

   = _____

2.

   = _____

3.

   = _____

4.

   = _____

5.

   = _____

69

Making Change

Write how much change is left over for each question.

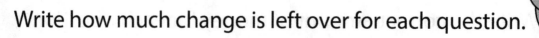

1.

 -  = _____

2.

 -   = _____

3.

 -   = _____

4.

 -  = _____

5.

 -  = _____

Time on a Clock

Write the time shown on the clock.

1.
 = ____

2. = ____

3.  = ____

4. = ____

5.  = ____

71

Time Matching

Match each Clock (on the left) with the correct Time (on the right).
Write your answer in the blank next to the clock.

1.  _____

2. 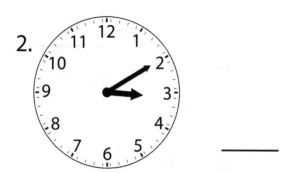 _____

3.  _____

4.  _____

5. 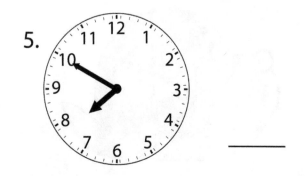 _____

A. 5:20

B. 5:10

C. 3:10

D. 11:40

E. 7:50

Passing of Time

Write the correct time on the line.

1. What time will it be in 1 hour and 15 minutes? _____

2. What time will it be in 3 hours and 5 minutes? _____

3. What time will it be in 2 hours and 40 minutes? _____

4. What time will it be in 35 minutes? _____

5. What time will it be in 2 hours and 25 minutes? _____

6. What time will it be in 4 hours and 45 minutes? _____

7. What time will it be in 35 minutes? _____

8. What time will it be in 3 hours and 10 minutes? _____

Answers: Chapter 7

Coin Value

1. 25 cents
2. 10 cents
3. 1 cents
4. 5 cents

Dollar Value

1. $10
2. $20
3. $1
4. $5

Counting Coins

1. 46 cents
2. 84 cents
3. $1.17
4. 87 cents
5. $1.80
6. 75 cents
7. $1.56
8. 58 cents

Counting Dollars

1. $35
2. $46
3. $88
4. $49
5. $83
6. $62
7. $97
8. $35

Counting Dollars and Coins

1. $32.65
2. $13.55
3. $34.41
4. $63.34
5. $41.53

Answers: Chapter 7

Making Change

1. 19 cents
2. 62 cents
3. 63 cents
4. $2.97
5. $2.10

Time Matching

1. D
2. C
3. B
4. A
5. E

Time on a Clock

1. 7:30
2. 11:40
3. 12:20
4. 9:50
5. 5:50

Passing Time

1. 8:15
2. 2:05
3. 1:10
4. 5:05
5. 11:15
6. 7:55
7. 5:55
8. 2:50

Chapter Eight: Geometry

Basic Shapes

How many smaller shapes can you find in each larger shape?
Write your answer in the blank.

1. 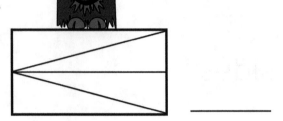 _____

2. 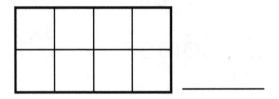 _____

3. 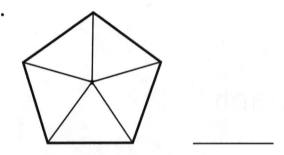 _____

4. 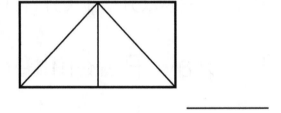 _____

5. 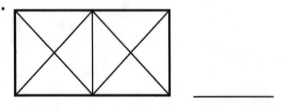 _____

6. 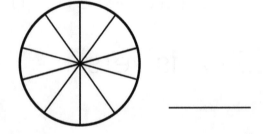 _____

7. 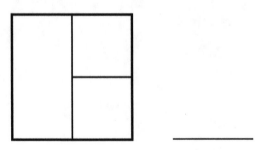 _____

8. 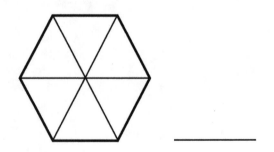 _____

9. 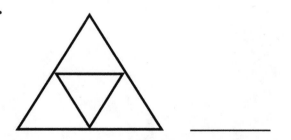 _____

10. 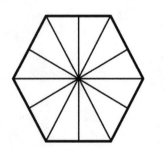 _____

Comparing Shapes by Sides

Write inside each shape how many sides it has.
Then, answer the question.

1.

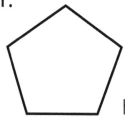 has how many more sides than 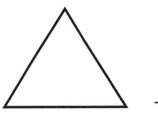 _____

2.

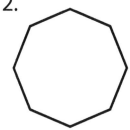 has how many more sides than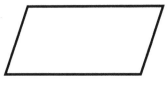

_____

3.

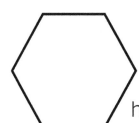 has how many more sides than  _____

4.

How many total sides do these shapes have?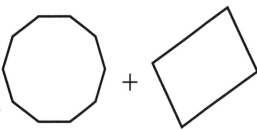

_____

5.

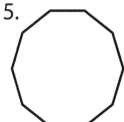 has how many more sides than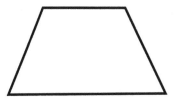

_____

Perimeter

Find the perimeter of each object by
adding all the sides. Write out the equations.
The first problem has been finished for you

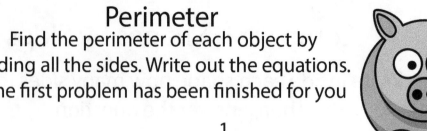

1.

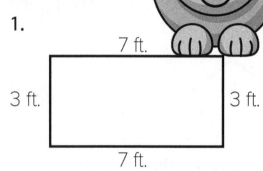

7 ft.

3 ft.   3 ft.

7 ft.

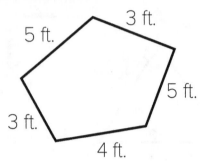

5 ft.   3 ft.

5 ft.

3 ft.

4 ft.

3 + 4 + 5 +3 + 5 = 20 ft.

_____

2.

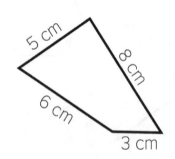

5 cm

8 cm

6 cm

3 cm

_____

3.

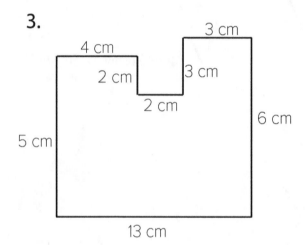

3 cm

4 cm

2 cm   3 cm

2 cm

5 cm   6 cm

13 cm

_____

4.

4 in.

5 in.

3 in.

2 in.

6 in.

5.

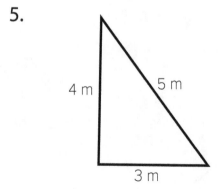

4 m   5 m

3 m

_____   _____

80

Area on a Graph
Find the area of each shape.

1. _____ units   4. _____ units

2. _____ units   5. _____ units

3. _____ units   6. _____ units

Drawing Perimeter on a Graph

Use the grid to draw the perimeter for each shape

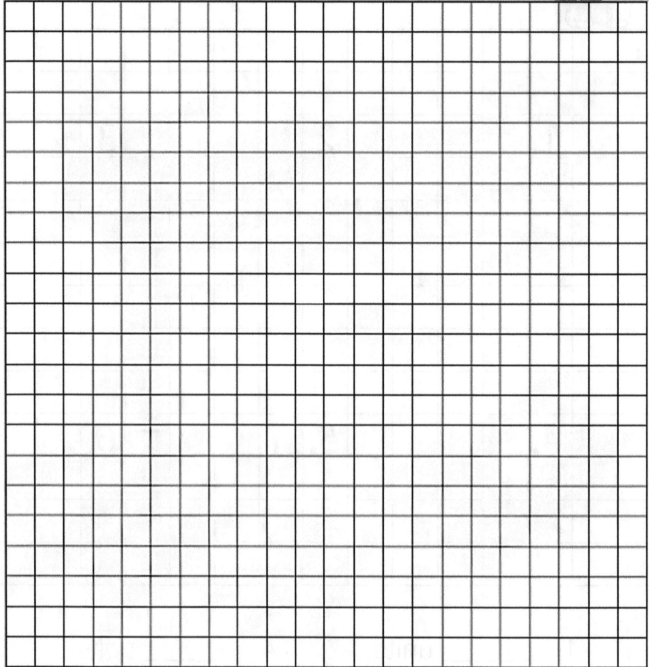

1. Draw 2 shapes with a perimeter of 10 units.

2. Draw 3 shapes with a perimeter of 14 units.

3. Draw 2 shapes with a perimeter of 16 units.

4. Draw 1 shape with a perimeter of 28 units.

5. Draw 2 shapes with a perimeter of 38 units.

Drawing Area on a Graph

Use the grid to draw the shapes for each question.

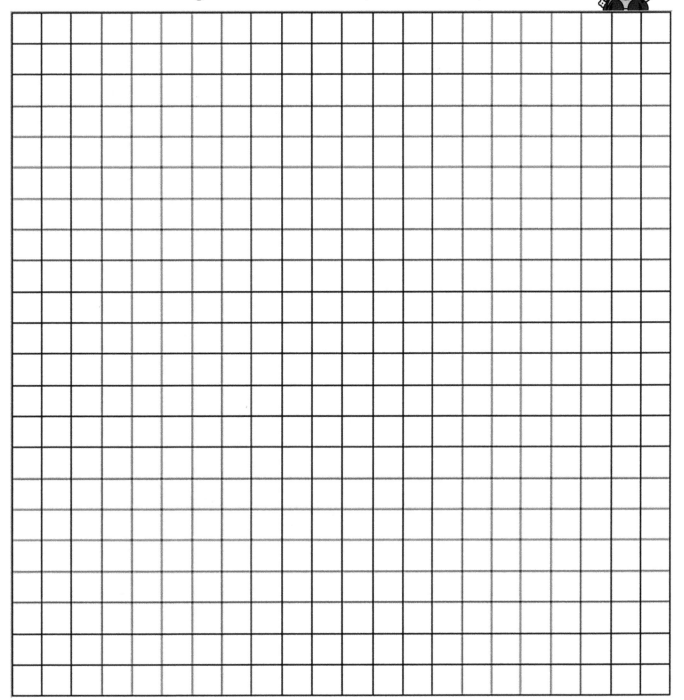

1. Draw 3 shapes with an area of 3 units each.

2. Draw 4 shapes with an area of 12 units each.

3. Draw 3 shapes with an area of 16 units each.

4. Draw 3 shapes with an area of 8 units each.

5. Draw 2 shapes with an area of 24 units each.

Answers: Chapter 8

Baic Shapes

1. 4
2. 6
3. 5
4. 4
5. 8
6. 10
7. 3
8. 6
9. 4
10. 12

Perimeter

1. 7 + 3 + 7 + 3 = 20 feet
2. 5 + 8 + 3 + 6 = 22 cm
3. 4 + 2 + 2 + 3 + 3 + 6 + 13 + 5 = 38 cm
4. 5 + 4 + 3 + 2 + 6 = 20 inches
5. 4 + 5 + 3 = 12m

Comparing Shapes by Sides

1. 2
2. 4
3. 1
4. 14
5. 6

Area on a Graph

1. 12
2. 11
3. 10
4. 9
5. 8
6. 6

Answers: Chapter 8

Drawing Perimeter on a Graph

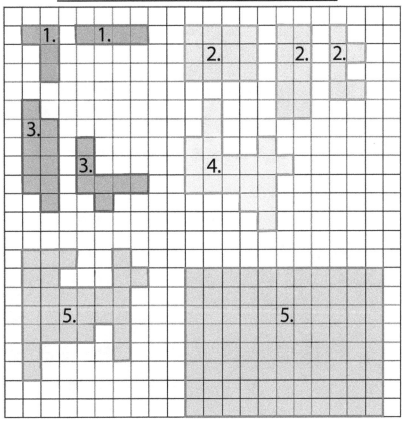

Drawing Area on a Graph

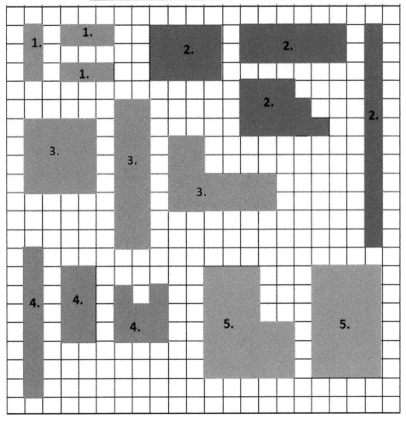

Chapter Nine: Graphs

Drawing on a Bar Graph

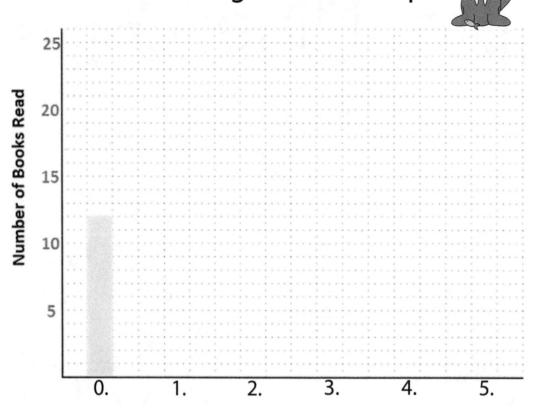

Use the Graph above to show the number of books each child read

0. Donna read 12
1. Ricky read 10 books.
2. Juan read 5 books.
3. Damien read 19 books.
4. Christina read 16 books.
5. Drew read 24 books.

Color the boxes in the graph below to show your answer.

6. How many dimes equal 6 nickels?
7. How many nickels equal 1 dime?
8. How many quarters equal $1.50?
9. How many dimes would you have left if you started with $1.00 worth of dimes and spent 60 cents?
10. How many dimes are equal to 2 quarters, 5 nickels, and 5 pennies?

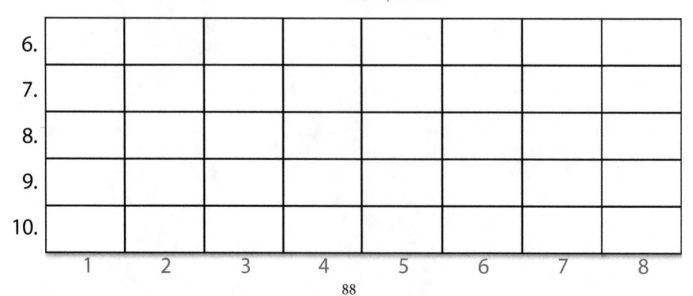

Reading a Line Graph
Use the graph to answer the questions

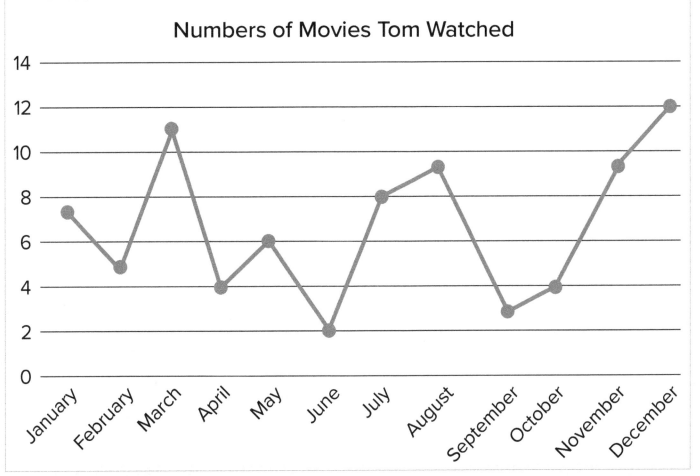

Numbers of Movies Tom Watched

1. In which month did Tom watch the most movies? _____

2. In which month did Tom watch the fewest movies? _____

3. How many movies did Tom watch in October? _____

4. How many movies did Tom watch in August? _____

5. How many more movies did Tom watch in January than he did in September? _____

6. How many fewer movies did Tom watch in October than he did in March? _____

7. How many movies did Tom watch in total from January through April (including both)? _____

8. Did Tom watch more movies in May or February? _____

9. Did Tom watch fewer movies in November or April? _____

10. Did the number of movies Tom watched increase or decrease from May to June? _____

Reading a Pie Chart

Use the pie charts showing the favorite desserts of students in
four third grade classrooms to answer the questions that follow.
Write your answer in the blank.

Miss Martha's Class

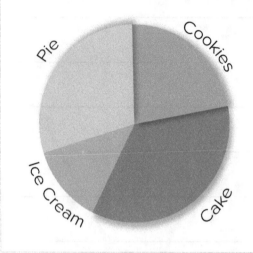

Mr. Bart's Class

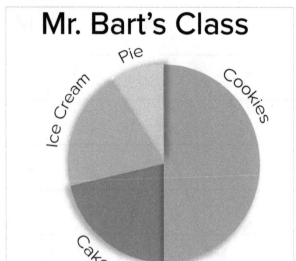

Miss Wendy's Class

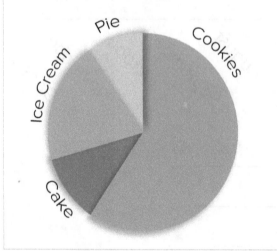

Mr. Peter's Class

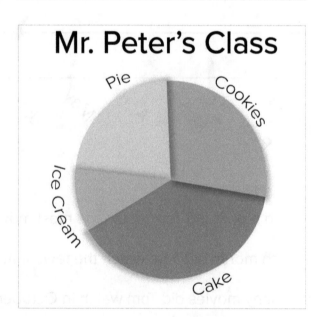

1. What dessert do most students in Miss Wendy's class prefer?

_____

2. In Mr. Bart's class, which dessert do the fewest students prefer?

_____

3. In Miss Martha's class, do more students like cake or cookies?

_____

4. In which teacher's class is pie a more popular favorite among students than it is in the other three classes?

_____

Shading a Bar Graph
Color a square for each object you see below.

	8				
7					
6					
5					
4					
3					
2					
1					

| Apples | Bananas | Strawberries | Oranges | Pears |

91

Word Problems and Graphs

Answer the following questions by completing the chart.

1. Carlos has done 40 pushups. If he does 15 pushups a day, how many pushups will he have done by the end of the week?

Carlo's Pushups	Monday	Tuesday	Wednesday	Thursday	Friday	Total

2. Grace spends $2.50 on school lunch every day. She's already spent 12 dollars. How much will she have spent by the end of the week?

Money Spent	Monday	Tuesday	Wednesday	Thursday	Friday	Total

3. Xavier practices cello for 20 minutes every day. He's already practiced 80 minutes. How many minutes will he have practiced in total at the end of the week?

Xavier's Minutes of Practice	Monday	Tuesday	Wednesday	Thursday	Friday	Total

4. Susan has drawn 14 pictures. If she draws 11 per day, how many drawings will she have by the end of the week?

Susan's Drawings	Monday	Tuesday	Wednesday	Thursday	Friday	Total

5. Victoria has 85 pieces of gum. If she chews 3 pieces per day, how many pieces will she have left at the end of the week?

Victoria's Pieces of Gum	Monday	Tuesday	Wednesday	Thursday	Friday	Total

Plotting Points on a Graph

Place the number in the correct location on the grid.

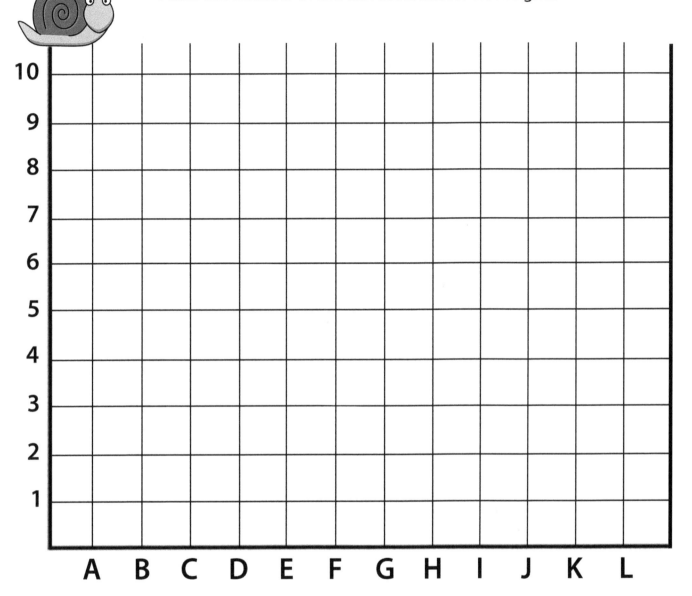

1. ① = A, 8   2. ② = C, 6

3. ③ = F, 4   4. ④ = I, 9

5. ⑤ = K, 2

Answers: Chapter 9

Drawing on a Bar Graph

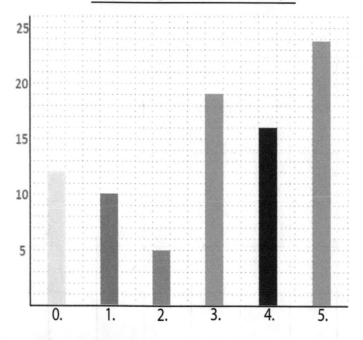

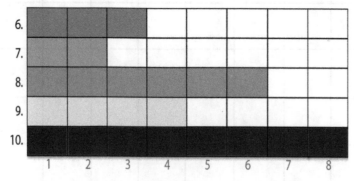

Reading a Line Graph

1. December
2. June
3. 4
4. 9
5. 4

6. 7
7. 27
8. May
9. April
10. Decrease

Reading a Pie Chart

1. Cookies
2. Pie
3. Cake
4. Miss Martha's class

Answers: Chapter 9

Shading a Bar Graph

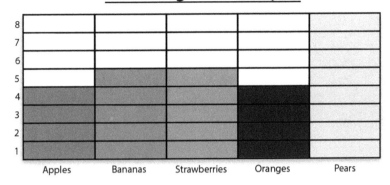

Word Problems and Graphs

1.

Carlo's Pushups	Monday	Tuesday	Wednesday	Thursday	Friday	Total
40	15	15	15	15	15	115

2.

Money Spent	Monday	Tuesday	Wednesday	Thursday	Friday	Total
12	2.50	2.50	2.50	2.50	2.50	24.50

3.

Xavier's Minutes of Practice	Monday	Tuesday	Wednesday	Thursday	Friday	Total
80	20	20	20	20	20	180

4.

Susan's Drawings	Monday	Tuesday	Wednesday	Thursday	Friday	Total
14	11	11	11	11	11	69

5.

Victoria's Pieces of Gum	Monday	Tuesday	Wednesday	Thursday	Friday	Total
85	3	3	3	3	3	70

Plotting Points on a Graph

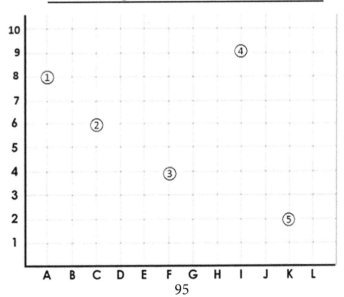

Chapter Ten: Word Problems

Addition Word Problems

1. Last year, Robin weighed 57 pounds. If her doctor tells her she has gained 8 pounds this year, how much does she now weigh?
   a. 63 pounds
   b. 64 pounds
   c. 65 pounds
   d. 66 pounds

2. Benny has 16 apples and picks 26 more.
   How many apples does he have in total?
   a. 32 apples
   b. 42 apples
   c. 38 apples
   d. 48 apples

3. Gwen has 38 rocks in her collection. If she gathers 16 more on a hike, how many rocks does she have?
   a. 54 rocks
   b. 56 rocks
   c. 62 rocks
   d. 64 rocks

4. Yoko has $3.80. If she earns 35 cents for vacuuming the kitchen, how much money will she have?
   a. $3.45
   b. $4.25
   c. $4.05
   d. $4.15

5. Henry has 59 comic books. If his grandmother buys him 13 more, how many will he have?
   a. 62 comic books
   b. 67 comic books
   c. 72 comic books
   d. 71 comic books

Subtraction Word Problems

1. Greg spent 12 dollars on pizza. He has 5 dollars left.
How much money did he start with?
a. $12
b. $17
c. $19
d. $21

2. Leah has to walk 32 miles to earn a prize. She has walked 18 miles.
How many more miles does she need to walk to earn the prize?
a. 20 miles
b. 18 miles
c. 8 miles
d. 14 miles

3. Emily has 21 shirts. If she donates 17 to charity, how many shirts does she have left?
a. 4 shirts
b. 17 shirts
c. 7 shirts
d. 5 shirts

4. Marcus has 24 candies left. If he started with 53 candies, how many did he eat?
a. 27
b. 29
c. 77
d. 19

5. Ricardo has $7.50 left. He spent $4.25 on pizza.
How much money did he start with?
a. $11.75
b. $11.25
c. $3.25
d. $3.75

Multiplication Word Problems

1. Sam has 6 friends. He wants to give each friend four drawings.
How many drawings does he need to make?
a. 24 drawings
b. 30 drawings
c. 20 drawings
d. 22 drawings

2. Thomas does 8 sit-ups every morning. How many sit-ups does he
complete after 5 days?
a. 35 sit-ups
b. 56 sit-ups
c. 48 sit-ups
d. 40 sit-ups

3. Juan has 3 chapters to read for each of the 6 classes he has.
How many chapters does he have to read in total?
a. 12 chapters
b. 18 chapters
c. 9 chapters
d. 21 chapters

4. Tara has 4 soccer practices each week. If the season is 9 weeks long,
how many practices will she have in total?
a. 39 practices
b. 32 practices
c. 19 practices
d. 36 practices

5. A teacher has 8 packs of markers. If each pack has 6 markers,
how many markers does she have in total?
a. 8 markers
b. 40 markers
c. 48 markers
d. 56 markers

Division Word Problems

1. Jeremy has 24 chocolate cookies that he evenly distributes among 8 friends. How many cookies does each friend receive?
a. 3 cookies
b. 4 cookies
c. 6 cookies
d. 12 cookies

2. Wendy has 7 days to read 56 pages in her book. If she wants to read the same number of pages each day, how many pages should she read per day?
a. 6 pages
b. 7 pages
c. 8 pages
d. 9 pages

3. Travis spent $45 on 5 movie tickets that were all equal in price. How much did each ticket cost?
a. $5
b. $9
c. $7
d. $11

4. Kerry is running laps around her school. If she ran 32 laps and her physical education teacher told her she completed 4 miles, how many laps equal 1 mile?
a. 4 laps
b. 8 laps
c. 12 laps
d. 6 laps

5. Paul has 36 pennies. He visits 6 fountains and throws in an equal number of pennies per fountain to make wishes. How many pennies does he throw into each fountain?
a. 3 pennies
b. 7 pennies
c. 8 pennies
d. 6 pennies

Answers: Chapter 10

Addition Word Problems

1. C
2. B
3. A
4. D
5. C

Subtraction Word Problems

1. B
2. D
3. A
4. B
5. A

Multiplication Word Problems

1. A
2. D
3. B
4. D
5. C

Division Word Problems

1. A
2. C
3. B
4. B
5. D

FREE Test Taking Tips DVD Offer

To help us better serve you, we have developed a Test Taking Tips DVD that we would like to give you for FREE. **This DVD covers world-class test taking tips that you can use to be even more successful when you are taking your test.**

All that we ask is that you email us your feedback about your study guide. Please let us know what you thought about it – whether that is good, bad or indifferent.

To get your **FREE Test Taking Tips DVD**, email freedvd@studyguideteam.com with "FREE DVD" in the subject line and the following information in the body of the email:

  a. The title of your study guide.

  b. Your product rating on a scale of 1-5, with 5 being the highest rating.

  c. Your feedback about the study guide. What did you think of it?

  d. Your full name and shipping address to send your free DVD.

If you have any questions or concerns, please don't hesitate to contact us at freedvd@studyguideteam.com.

Thanks again!

CPSIA information can be obtained
at www.ICGtesting.com
Printed in the USA
BVHW050401120221
599826BV00012B/821